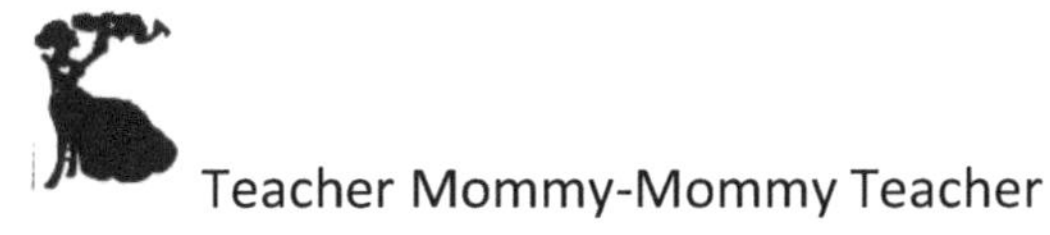

Images Credit

Rabbit, Log, Quarter &Teddy Bear Image by OpenClipart-Vectors from Pixabay

Squid, Dog and Cat Image by OpenClipart-Vectors from Pixabay

Teddy Bear with Pink Shirt Image by simisi1 from Pixabay

Lamb Image by bknis from Pixabay

Dollar Image by Mary Pahlke from Pixabay

Penny Image by b0red from Pixabay

Nickel & Dime clipart-library.com

Kids Behind the door, Sphere, Cube, Cylinder & Cone www.pinclipart.com

Pyramid www.clipartkey.com

Hippo, Squid, Girl with dog & Rain Cloud Image by Clker-Free-Vector-Images from Pixabay

Girl and Boy Teddy Bear Image by Laura Clayton from Pixabay

Feather Image by gdakaska from Pixabay

KINDERGARTEN MATH PROCESS CHARTS

Table of Content

Numbers

Number	Number Names	Count
1	one	
2	two	
3	three	
4	four	
5	five	
6	six	

Numbers

Numbers	Number Names	Count
7	seven	
8	eight	
9	nine	
10	ten	
11	eleven	
12	twelve	

Numbers

Number	Number Names	Count
13	thirteen	
14	fourteen	
15	fifteen	
16	sixteen	

Numbers

Number	Number Names	Count
17	seventeen	
18	eighteen	
19	nineteen	
20	twenty	

Numbers 1-50

1	2	3	4	5
6	7	8	9	10
11	12	13	14	15
16	17	18	19	20
21	22	23	24	25
26	27	28	29	30
31	32	33	34	35
36	37	38	39	40
41	42	43	44	45
46	47	48	48	50

Hundreds Chart

1	2	3	4	5	6	7	8	9	10
11	12	13	14	15	16	17	18	19	20
21	22	23	24	25	26	27	28	29	30
31	32	33	34	35	36	37	38	39	40
41	42	43	44	45	46	47	48	49	50
51	52	53	54	55	56	57	58	59	60
61	62	63	64	65	66	67	68	69	70
71	72	73	74	75	76	77	78	79	80
81	82	83	84	85	86	87	88	89	90
91	92	93	94	95	96	97	98	99	100

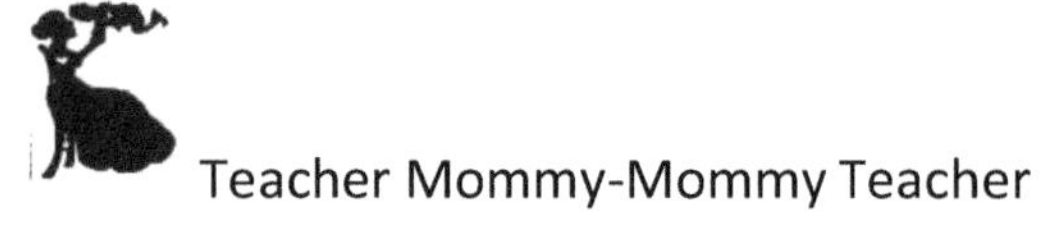

Ten Frames

Use the ten frames to add or subtract numbers less than or equal to ten. Counters can be arranged in different ways to represent a number.

Ten Frames

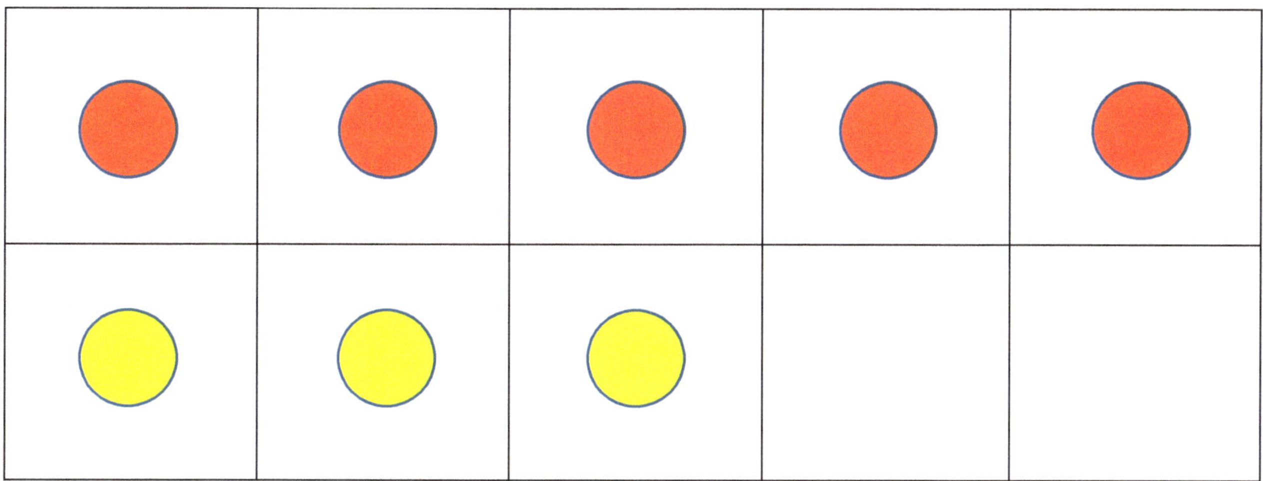

Examples:

5+3=8

1. Place five counters in the top row.
2. Place three counters in the bottom row.
3. Add all the counters.

How many in all?

Ten Frames

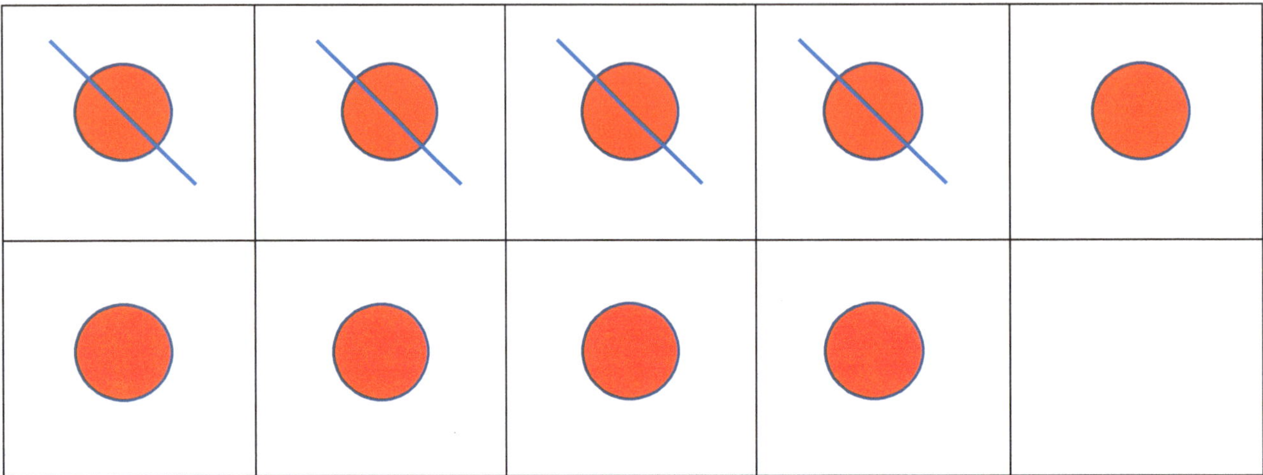

Examples:

9-4=5

1. Place nine counters in the frame.
2. Remove four of the counters.
3. Count how many are left.

How many are left?

How Many?

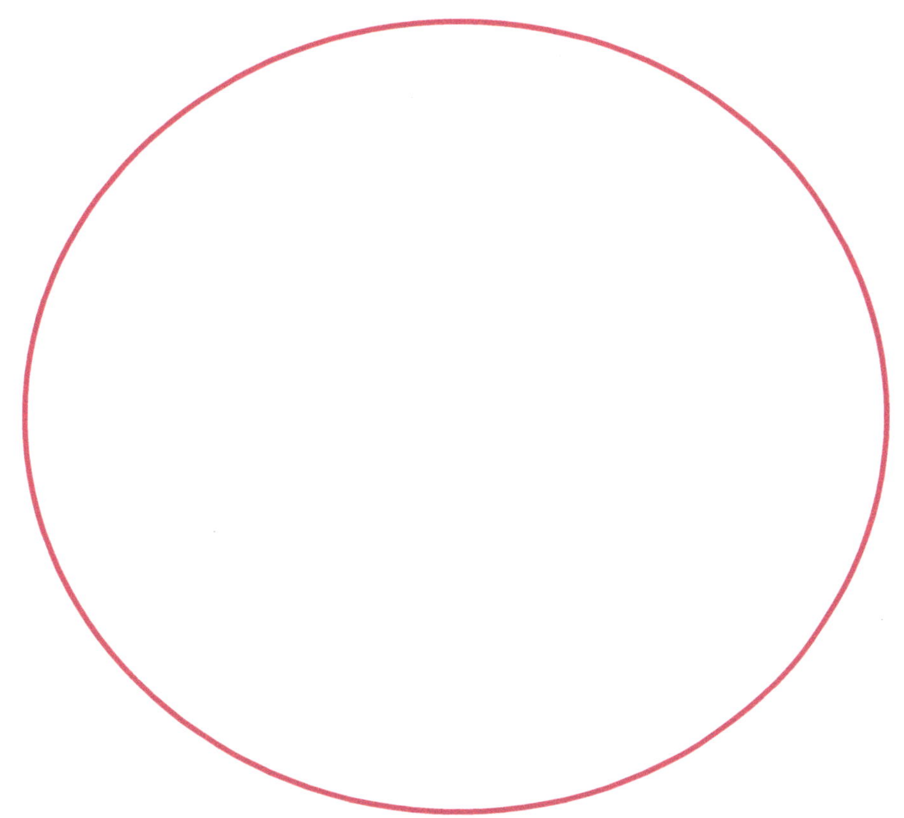

Use the circle to add or subtract numbers less than or equal to twenty.

How Many?

Examples:

10+3=13

1. Place ten objects in the circle.
2. Place three more objects in the circle.
3. Add all the objects.

How many in all?

How Many?

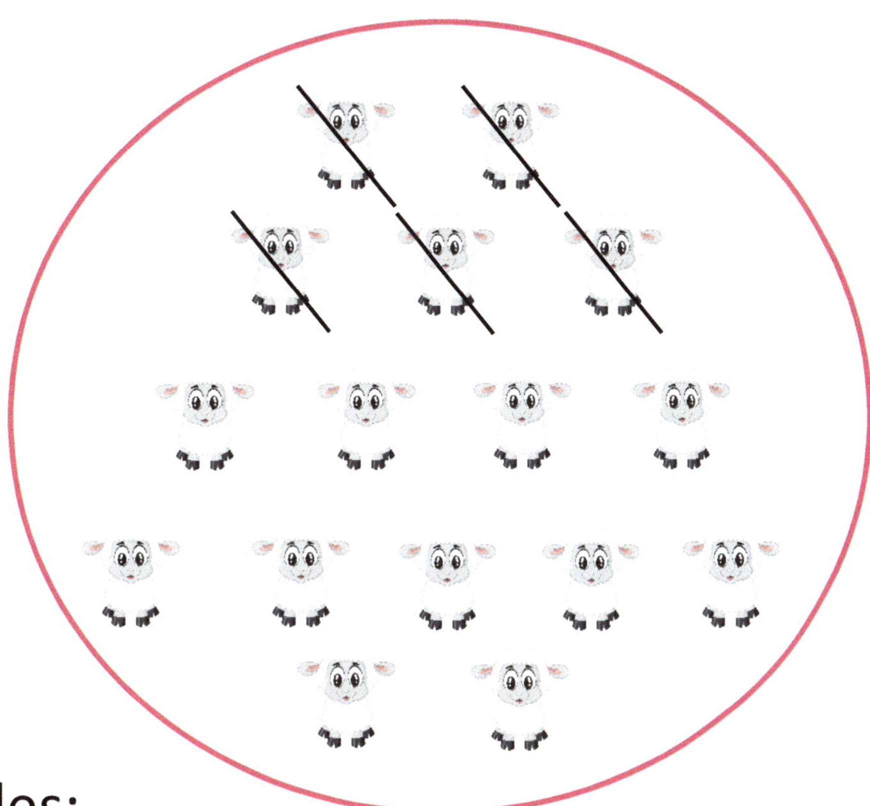

Examples:

16-5=11

1. Place sixteen objects in the circle.
2. Remove five from the circle.
3. Count how many are left.

How many are left?

Compare Numbers

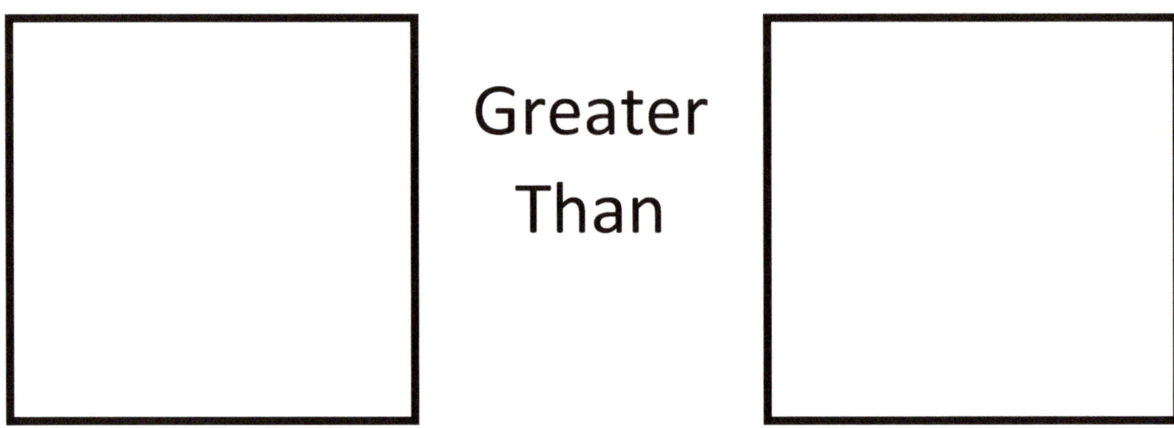

Greater than means, it has the largest amount. Place objects in the square to see which has more.

Greater Than

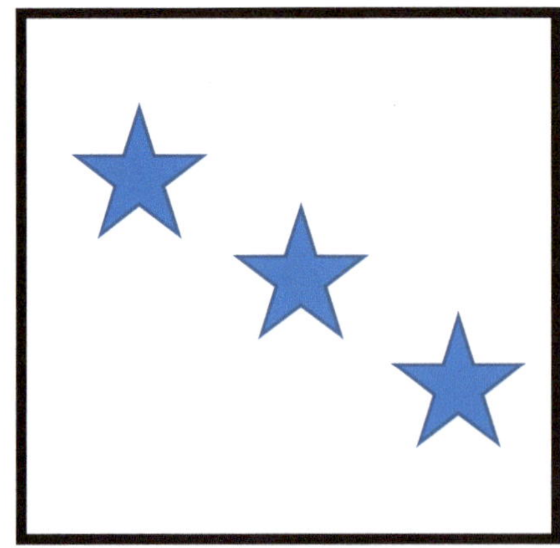

Compare Numbers

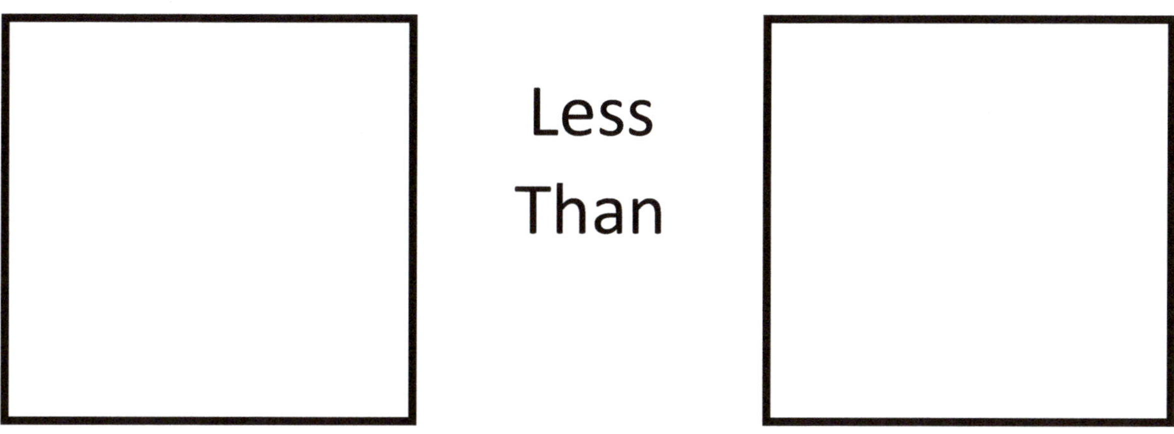

Less than means, it has the smallest amount. Place objects in the square to see which has less.

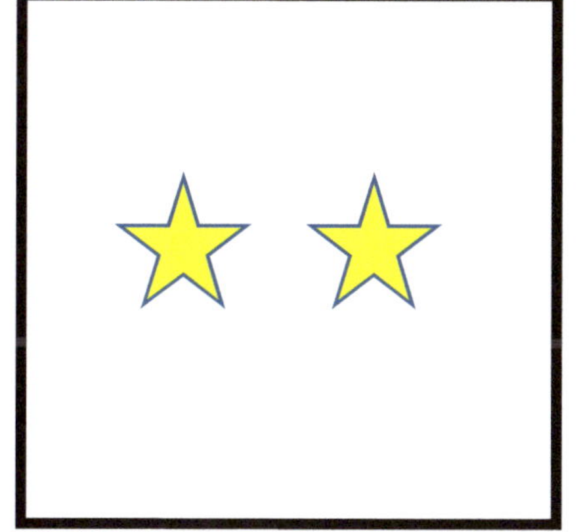

Less Than

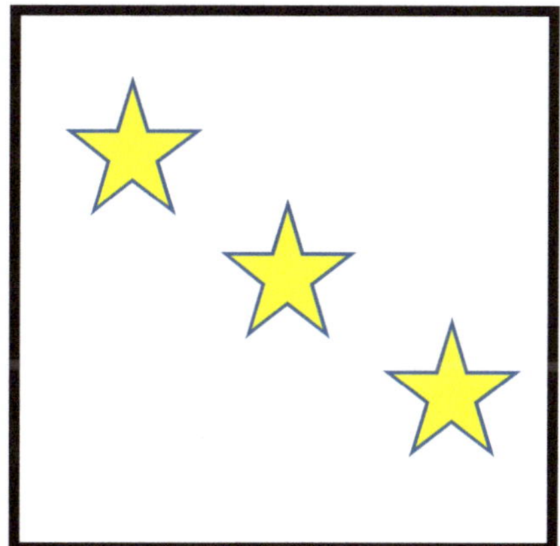

Compare Numbers

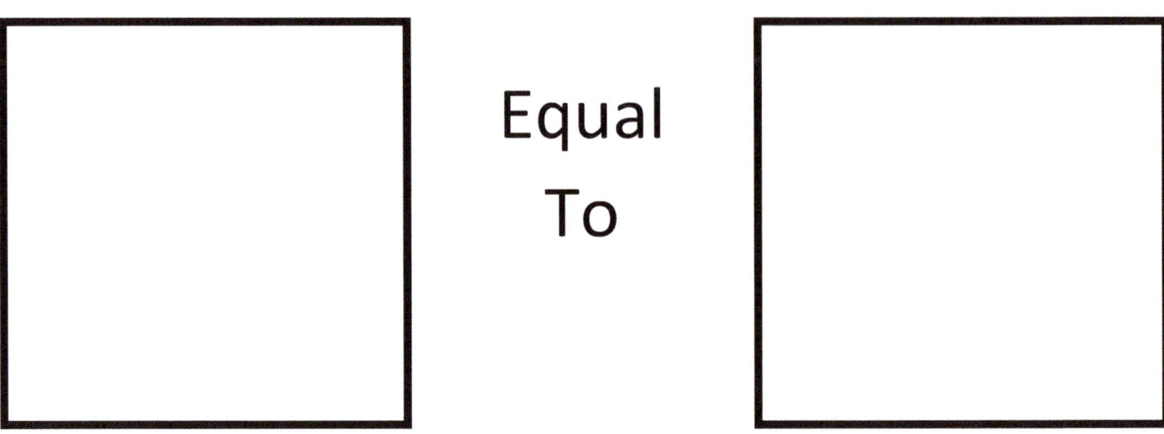

Equal to means, it has the same amount. Place objects in the square to see if they are the same.

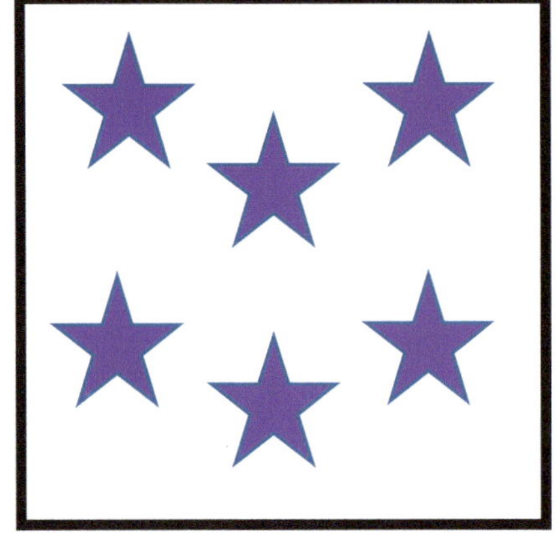

Equal To

Part-Part-Whole (Number Bonds)

Addition is putting together and adding to.

Subtraction is taking apart and taking from.

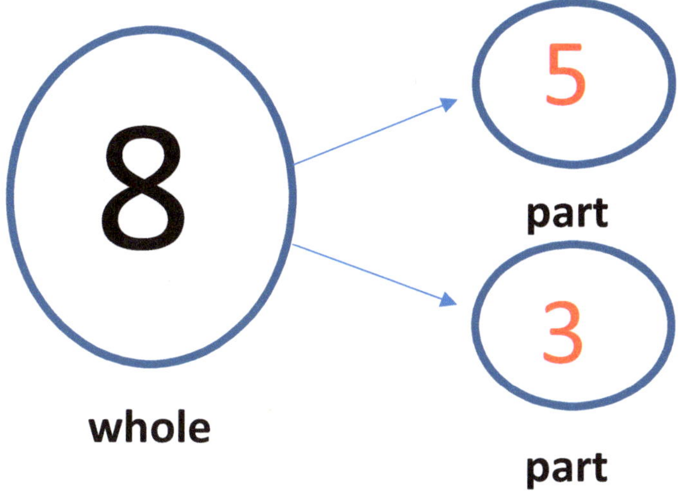

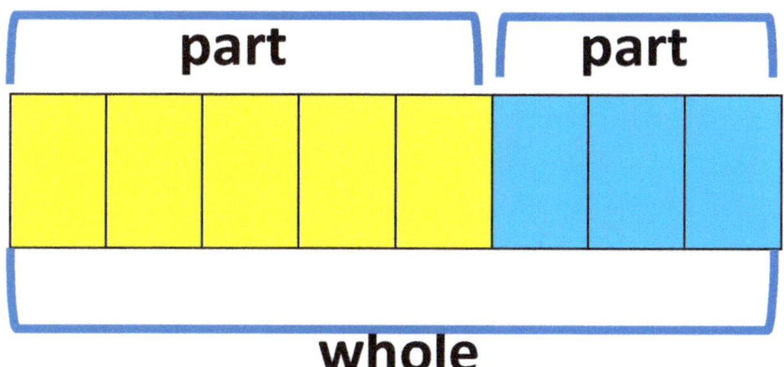

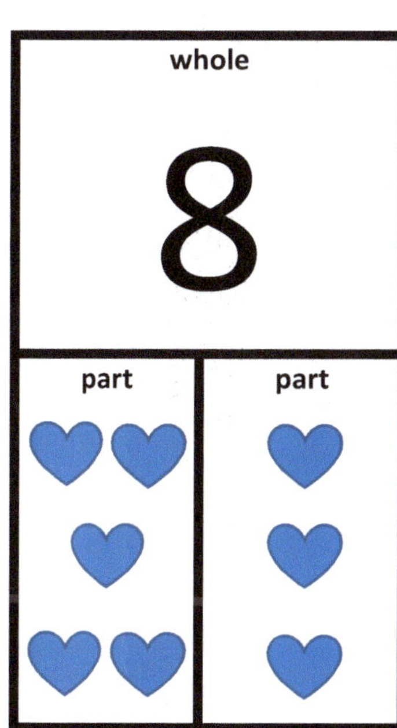

5+3=8

part + part = whole

8-5=3

whole – part = part

Adding using Number Lines

Place your finger on the first addend. Then jump forwards the second addend. The sum will be where your finger lands once you finish jumping.

5+3=8

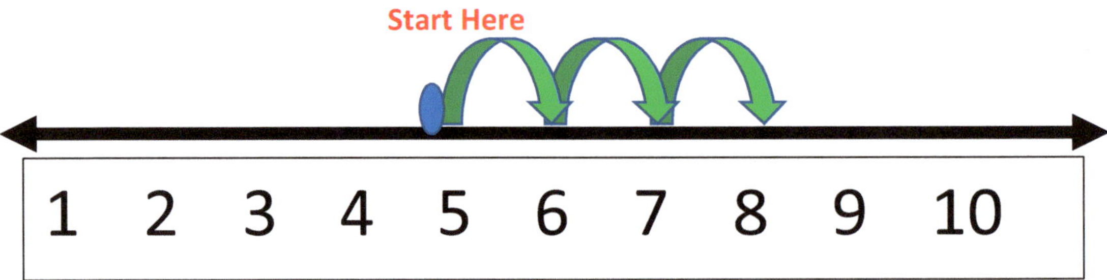

Subtracting using Number Lines

Place your finger on the first subtrahend. Then jump backwards towards the second subtrahend. The difference will be where your finger lands once you finish jumping.

8-5=3

Word Problem Strategy

C- Circle Key Words

(What do I know?)

U- Underline the Question

(What am I being asked to do?)

B- Box Any Clue Words

(Am I going to add, subtract, multiply or divide?)

E- Evaluate and Eliminate

(What steps do I take? What information don't I need?)

S- Show and Check Your Work

(Did I answer the question? How can I double check?)

Word Problems Vocabulary

Add

plus sum and total

in all altogether add

Subtract

left fewer minus

less take subtract

difference

Word Problems

7 apples are in the tree.
2 apples are on the ground.
<u>How many apples in all?</u>

$$7+2=9$$

Teacher Mommy-Mommy Teacher

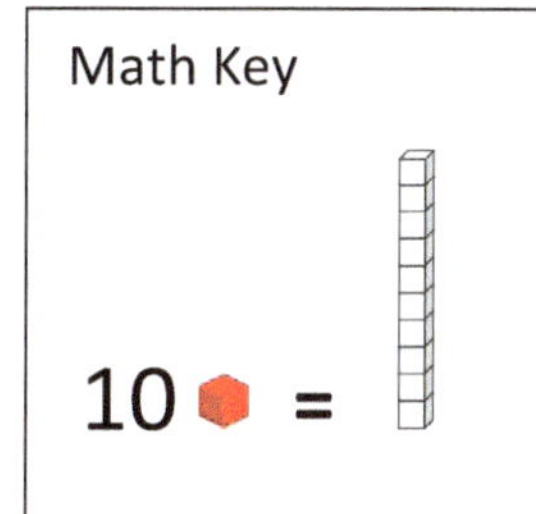

Place Value

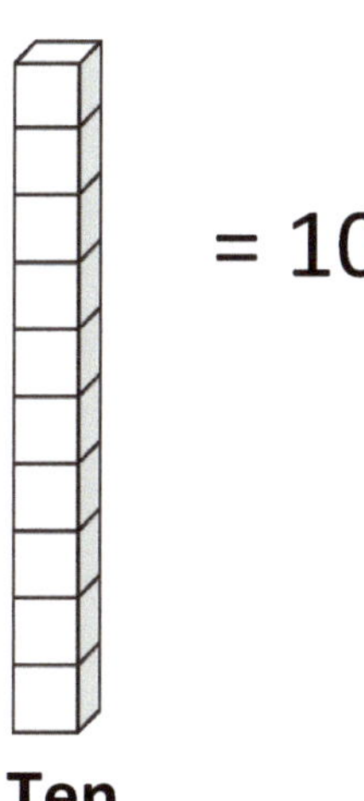

= 10

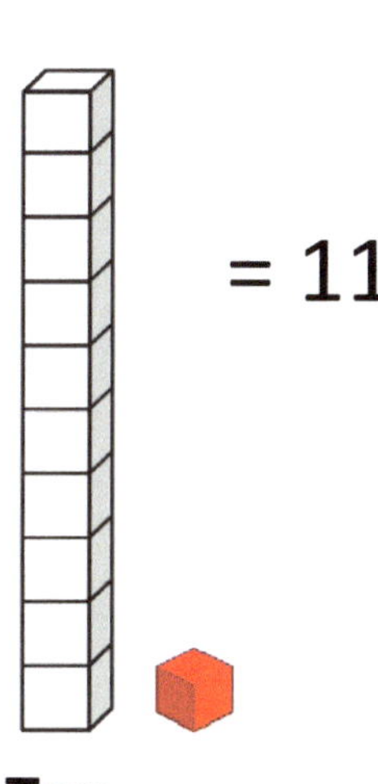

= 11

Ten one

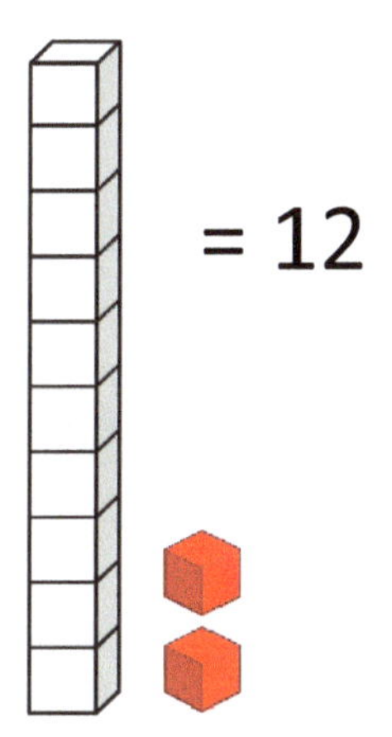

= 12

Ten ones

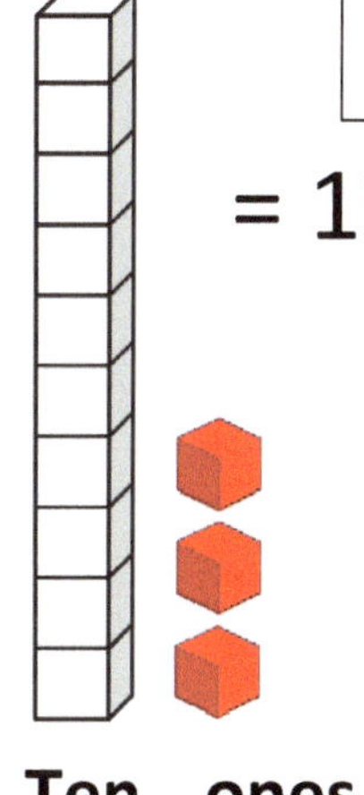

= 13

Ten ones

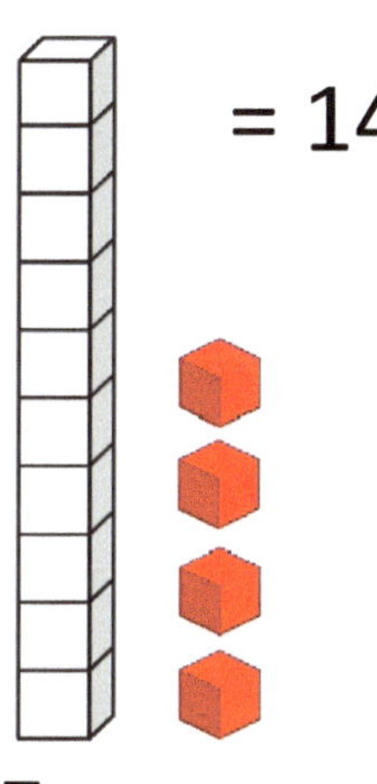

= 14

Ten ones

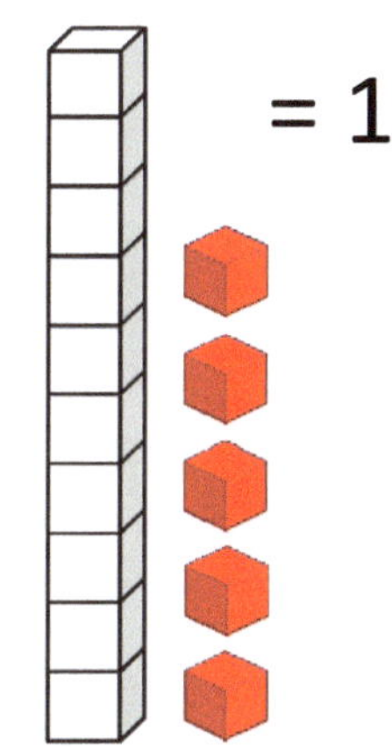

= 15

Ten ones

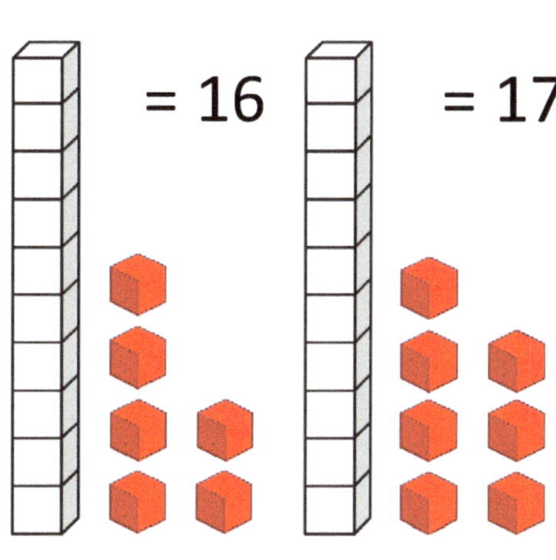

= 16

Ten ones

= 17

Ten ones

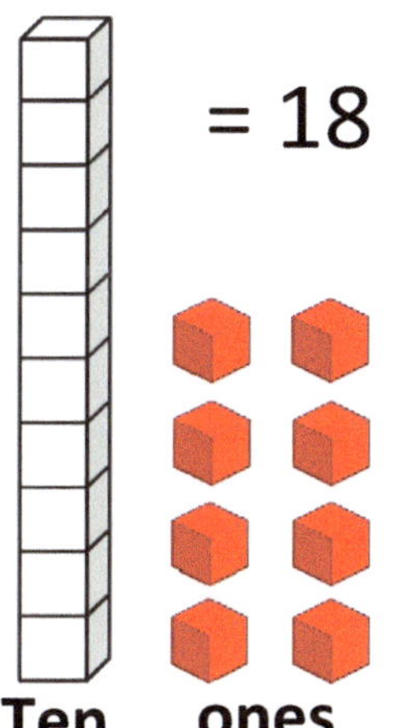

= 18

Ten ones

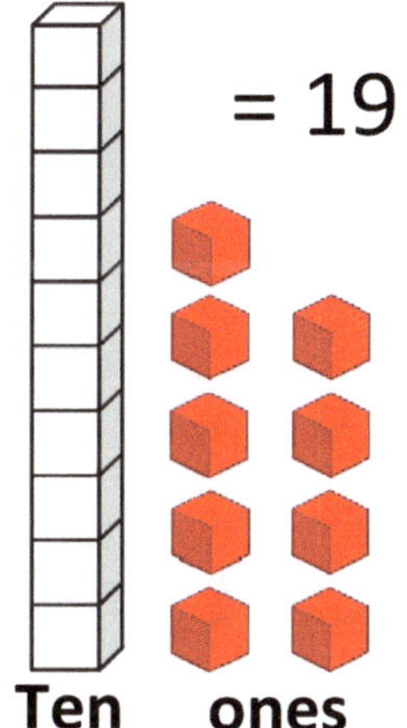

= 19

Ten ones

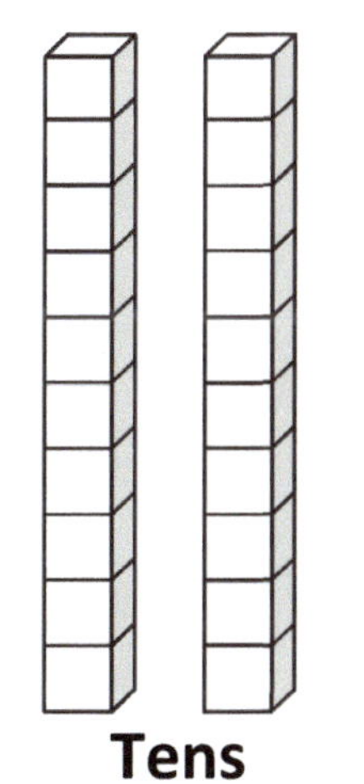

= 20

Tens

Months of the Year

January
February
March
April
May
June
July
August
September
October
November
December

Days of the Week

Sunday
 Monday
 Tuesday
 Wednesday
 Thursday
 Friday
 Saturday

Telling Time

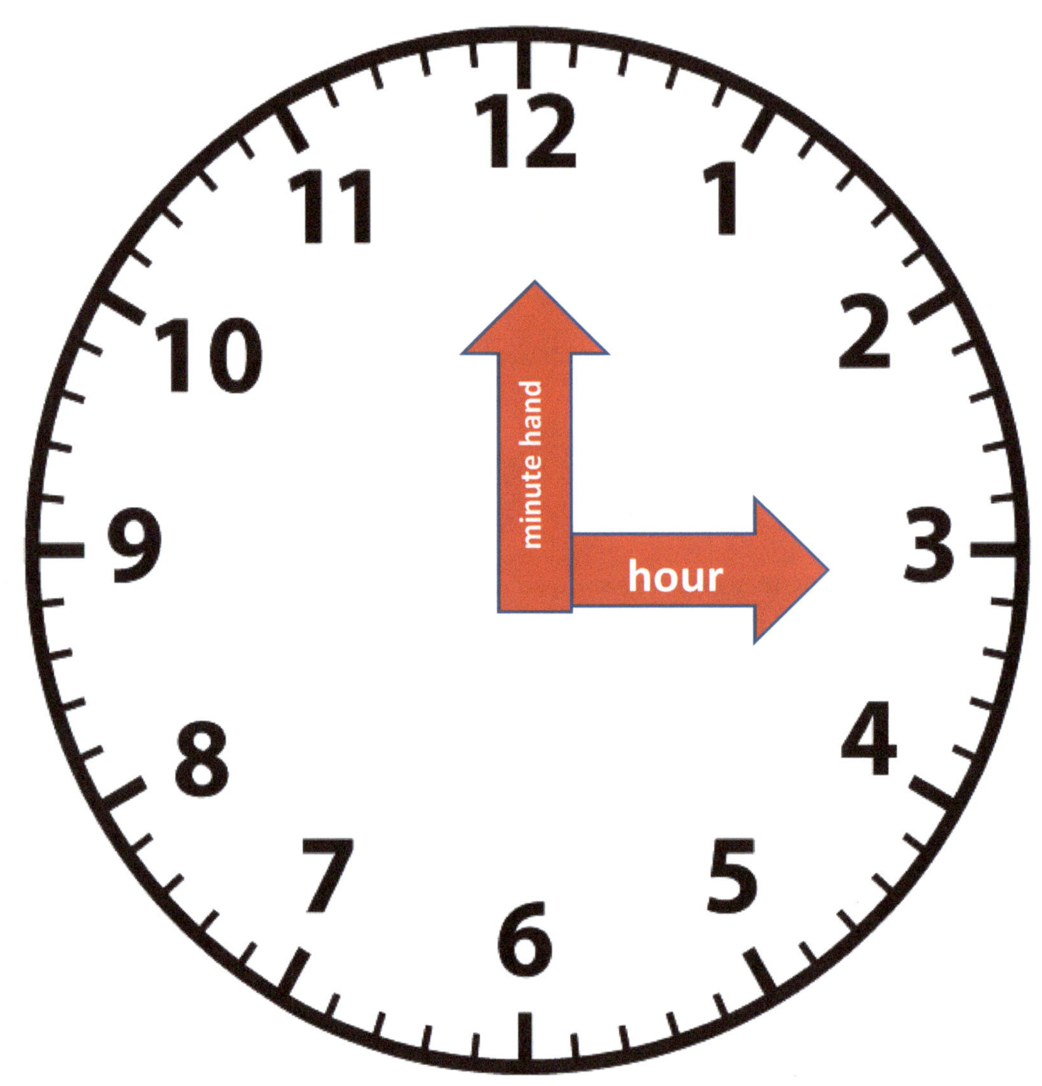

Calendar

August						
Sunday	Monday	Tuesday	Wednesday	Thursday	Friday	Saturday
						1
2	3	4	5	6	7	8
9	10	11	12	13	14	15
16	17	18	19	20	21	22
23	24	25	26	27	28	29
30	31					

2 Dimensional Shapes

2 dimensional shapes are **flat plane figures**.

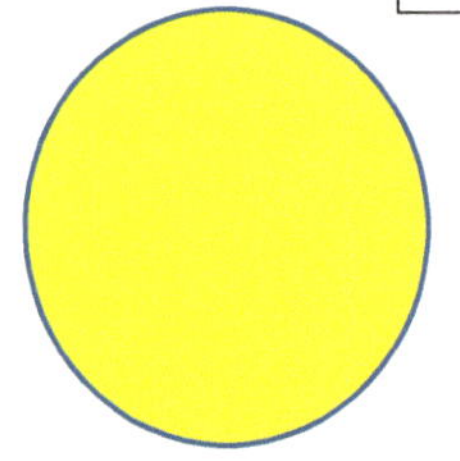

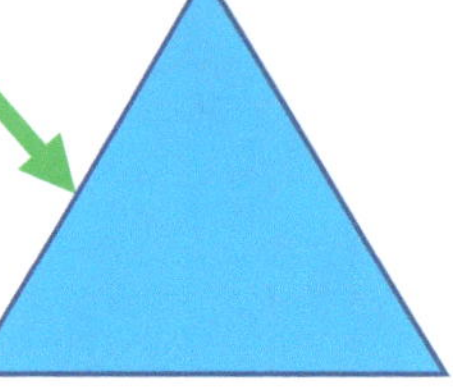

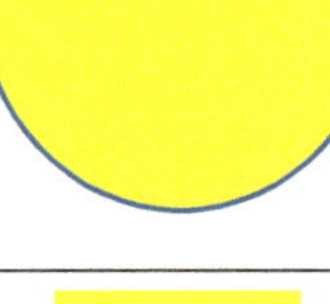

Circle	Triangle	Rectangle
0 sides	3 sides	4 sides

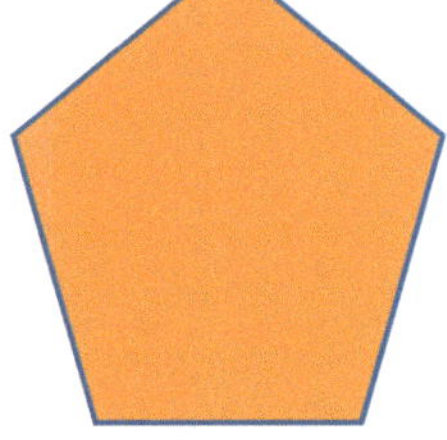

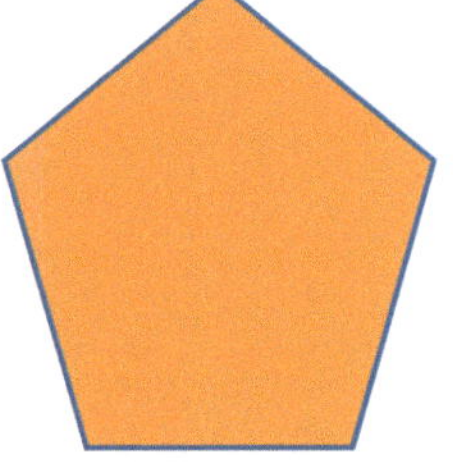

Diamond/ Rhombus	Square	Pentagon
4 sides	4 sides	5 sides

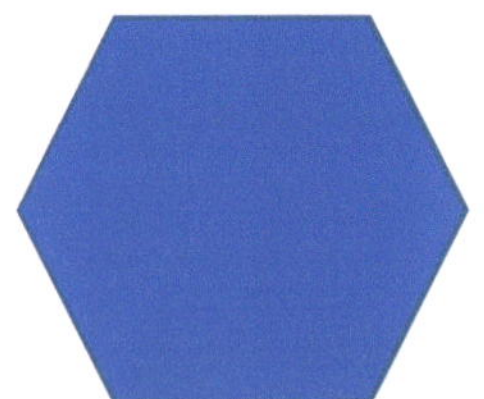

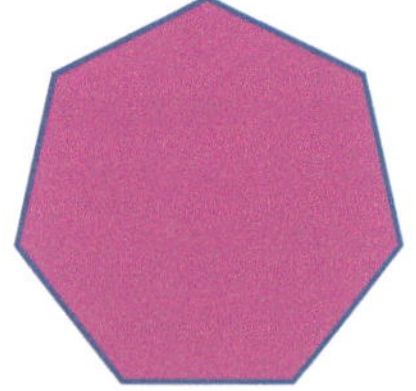

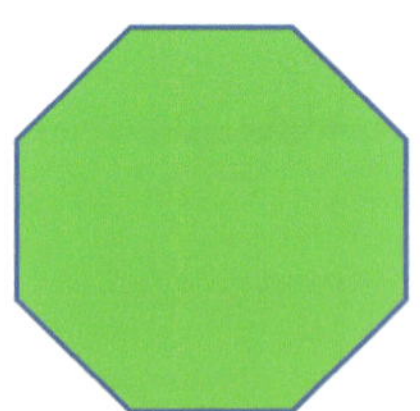

Hexagon	Heptagon	Octagon
6 sides	7 sides	8 sides

3 Dimensional Shapes

3 dimensional shapes are solid figures with faces, edges and vertices.

Cylinder

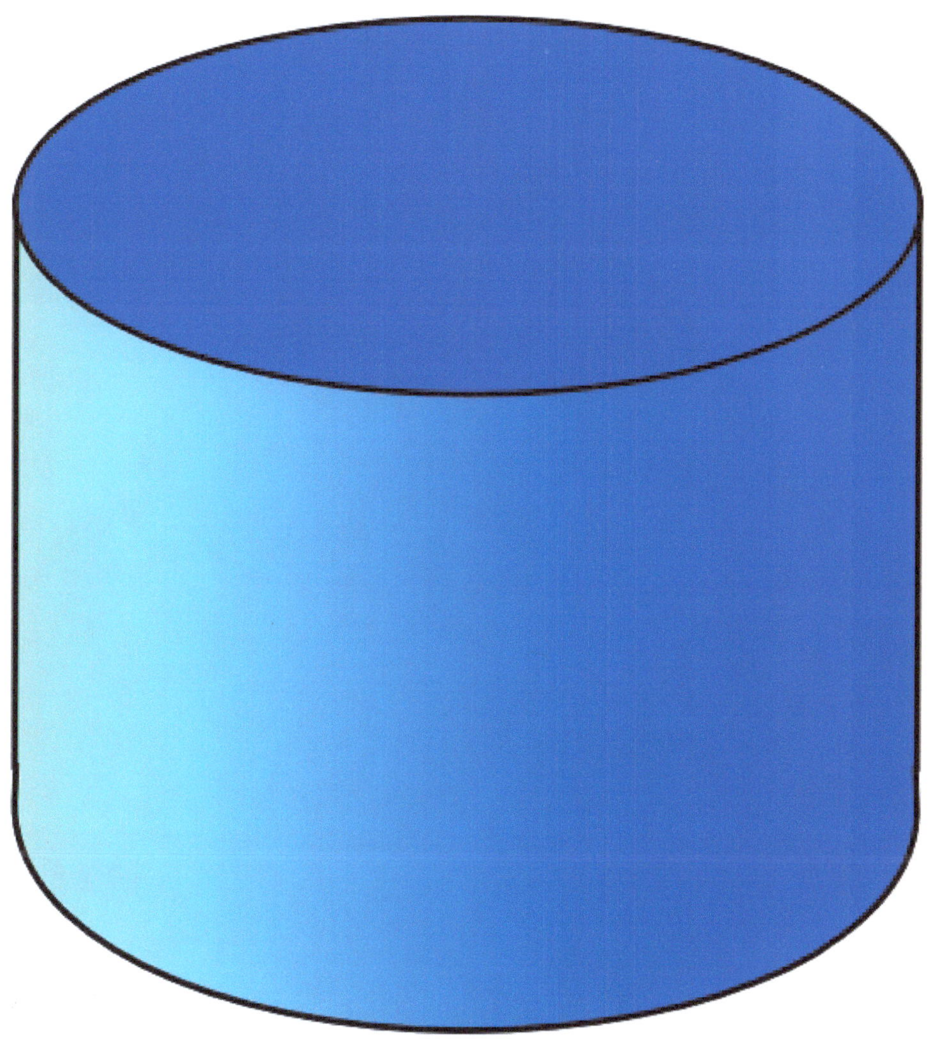

3 Dimensional Shapes

3 dimensional shapes are **solid figures with faces, edges and vertices.**

Pyramid

3 Dimensional Shapes

3 dimensional shapes are solid figures with faces, edges and vertices.

Cone

3 Dimensional Shapes

3 dimensional shapes are **solid figures with faces, edges and vertices.**

Cube

3 Dimensional Shapes

3 dimensional shapes are solid figures with faces, edges and vertices.

Sphere

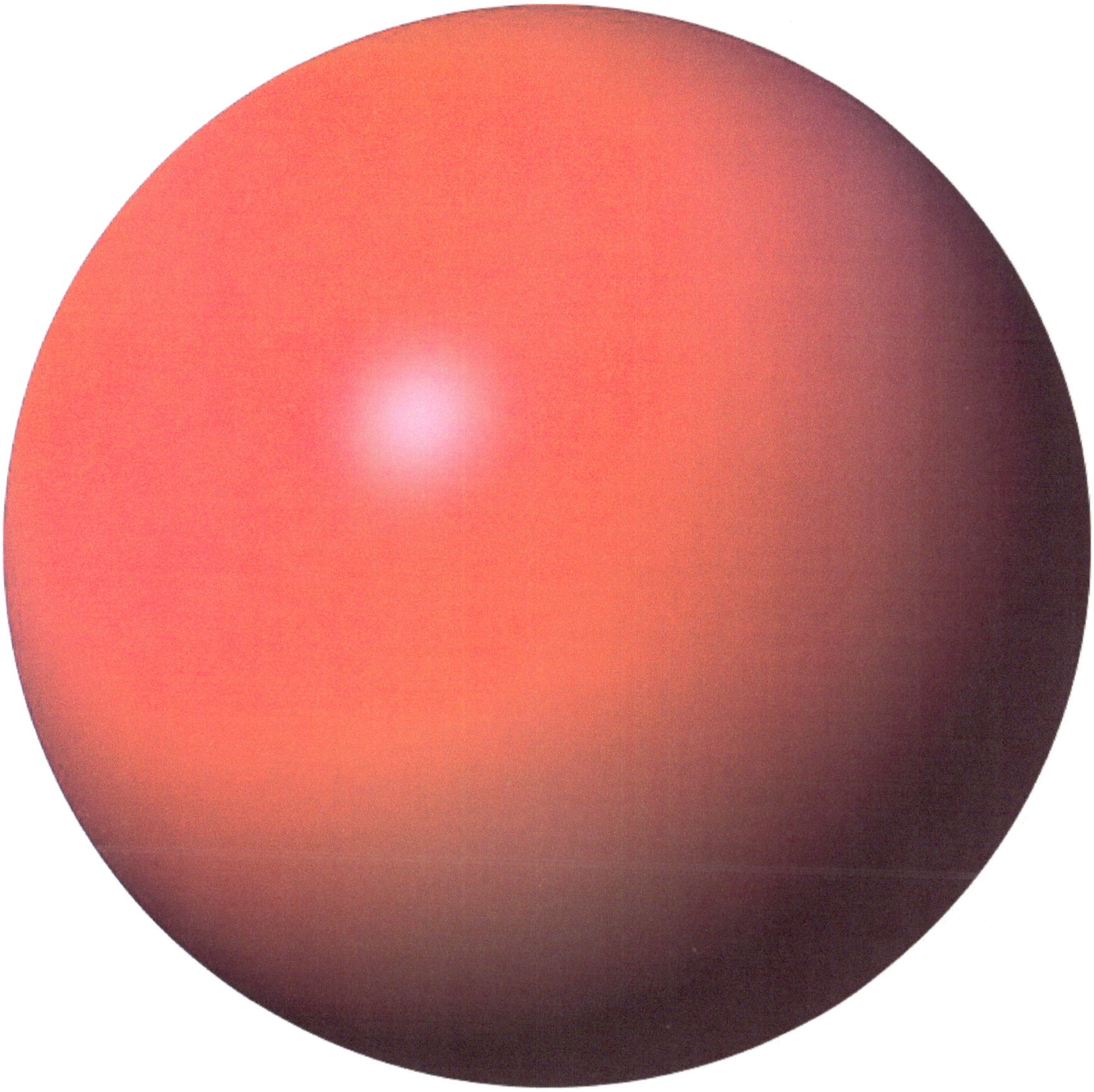

Money

Money is a **currency that is accepted as payment for goods and services.**

Penny

1¢

Money

Money is a **currency that is accepted as payment for goods and services.**

Nickel

5¢

Money

Money is a currency that is accepted as payment for goods and services.

Dime

10¢

Money

Money is a currency that is accepted as payment for goods and services.

Quarter

25¢

Money

Money is a currency that is accepted as payment for goods and services.

Dollar

$1.00

Light or Heavy

More or Less

Short or Tall

Above

Below

Beside

In Front Of

Behind

Next to

www.ingramcontent.com/pod-product-compliance
Lightning Source LLC
LaVergne TN
LVHW070154110826
845147LV00002B/397
9781734016871